DISCARD

This is the story of one of the great revolutions in the life history of man, a time when he first learned to grow crops and domesticate animals. Before, he had been a seed gatherer, wandering from place to place in search of food, at the mercy of his environment. Now, for the first time, he could settle in one place. He developed crafts—basketry, weaving, and pottery—and conceived the idea of property.

A striking pictorial treatment of the Neolithic Age—farming methods, tools and crafts, art, and religion. Leonard Weisgard's pictures convey with strength and precision the character of man's life in this early time and that dramatic boldness found in all the first farmers' design and artifacts. These ancient tools and weapons are lasting evidence of the creativity which enabled early man to play an active part in determining his own environment.

THE FIRST FARMERS is the story of a great step forward in man's long movement toward civilization.

LIFE LONG AGO

Editorial and Historical Consultant **Edward Ochsenschlager**
Archaeologist and Lecturer in Classics

Education Consultant **Rosemary Daly**
Librarian, Ethical Culture School,

THE FIRST FARMERS

IN THE NEW STONE AGE

LEONARD WEISGARD

COWARD-McCANN
New York

The first farmers who are our cultural ancestors almost as long ago as 11,000 years lived in the settlements in the Near East indicated on the map on page 10. Neolithic communities grew up in many other places. In this book pictures and text describe the revolution which took place in the Near East. Any material found in other regions will be identified.

Today when out of this Neolithic culture our highly developed technological civilization has evolved, it is important to know that man in regions as far apart as New Guinea and remote parts of Peru, Brazil and Hawaii is holding to Neolithic beliefs. He is still using stone tools, building homes in a Neolithic fashion, obtaining and preparing foods in a Neolithic way. These are truly Neolithic peoples living in the twentieth century.

0110 UP
© 1966 by Leonard Weisgard. All rights reserved.
Library of Congress Catalog Card Number: 65-20381
Printed in the United States of America

It is, of course, pre-written history that we mean when we speak of prehistory.... The material which the prehistorian uses—which he interprets—is, by definition, and by fact, unwritten. It is the unwritten remains of the early past of man, the mute, silent witness of the origins and early development of prehistory—tools, weapons, houses, temples, paintings, tombs, farms, fields, forts, water mills—in a word it is all archaeological material.
  Glyn Daniel, *The Idea of Prehistory*

Without Braidwood, Kenyon, Mellaart, Ochsenschlager, Singer, Alice Torrey and others listed in the back, *The First Farmers* could never have been put together into this book. I thank them one and all.

CONTENTS

How We Know What We Know	9
The First Farmer	11
Where Man Farmed	13
Early Food Crops	14
How Man Farmed	15
Man's Uses of Plants	17
Man Domesticates Animals	19
Tools	22
Homes	26
Basketry	30
Spinning and Weaving	32
Clothing	34
Transportation	36
Pottery	40
Art	46
Religion	50
Conclusion	54
MAP OF IMPORTANT EXCAVATIONS	56
IMPORTANT SELECTED NEOLITHIC SITES	57
GLOSSARY AND PRONUNCIATION GUIDE	58
LIST OF BOOKS FOR FURTHER READING	59
INDEX	60

THE FIRST FARMER'S PLACE IN TIME

1770 A.D. — INDUSTRIALIZATION

CIVILIZATION — 3000 B.C. — FOOD PRODUCTION

Farming Begins — NEOLITHIC AGE — 7000

MESOLITHIC — 10,000

UPPER PALEOLITHIC — 35,000

50,000

MIDDLE PALEOLITHIC — 100,000

150,000

350,000

400,000

550,000

Beginning of Tool Using

LOWER PALEOLITHIC — 800,000

1,750,000

HOW WE KNOW WHAT WE KNOW

This book is about one period of prehistory, a period known as the Neolithic or New Stone Age. It began about 9,000 years ago. There are no written records of it. Man had not yet learned to write. This story has been built up from clues, from the tangible, visible remains that early man left behind him—the bones, tools, weapons, fragments of utensils, remnants of houses. Many kinds of scientists have studied these clues—archaeologists, geologists, astronomers, biologists, naturalists, anthropologists. By piecing together the evidence, they have reconstructed as nearly as they can, the life of early man.

We may never be able to name them, but this story in a very exciting way is a record of discoverers and inventors. These people were the revolutionaries. Their discovery of how to plant and harvest seeds, of how to domesticate animals, brought about a revolution in man's way of life which made possible everything that has happened since. It freed man from the need to give all of his time to the necessary, to just staying alive. Now he could begin to create and to give form to a world of ideas and imaginings. In its deceptively simple way this time was the basis, the beginning of all wealth, all leisure and all art.

The clues have all been buried during the centuries between then and now. Once the place where they are has been located and the archaeologist with the utmost care has dug them up, the business of prime importance begins. What do they mean? What can they tell us?

Scientists may be able to draw only tentative conclusions from all this evidence, but they are able to date what they find quite accurately.

By measuring the amount of radioactivity remaining in organic substances, particularly in carbon material—charcoal and charred bones—a scientist is able to determine its approximate age up to 50,000 years. Radioactive matter loses its radioactive count at a steady rate with the passage of time.

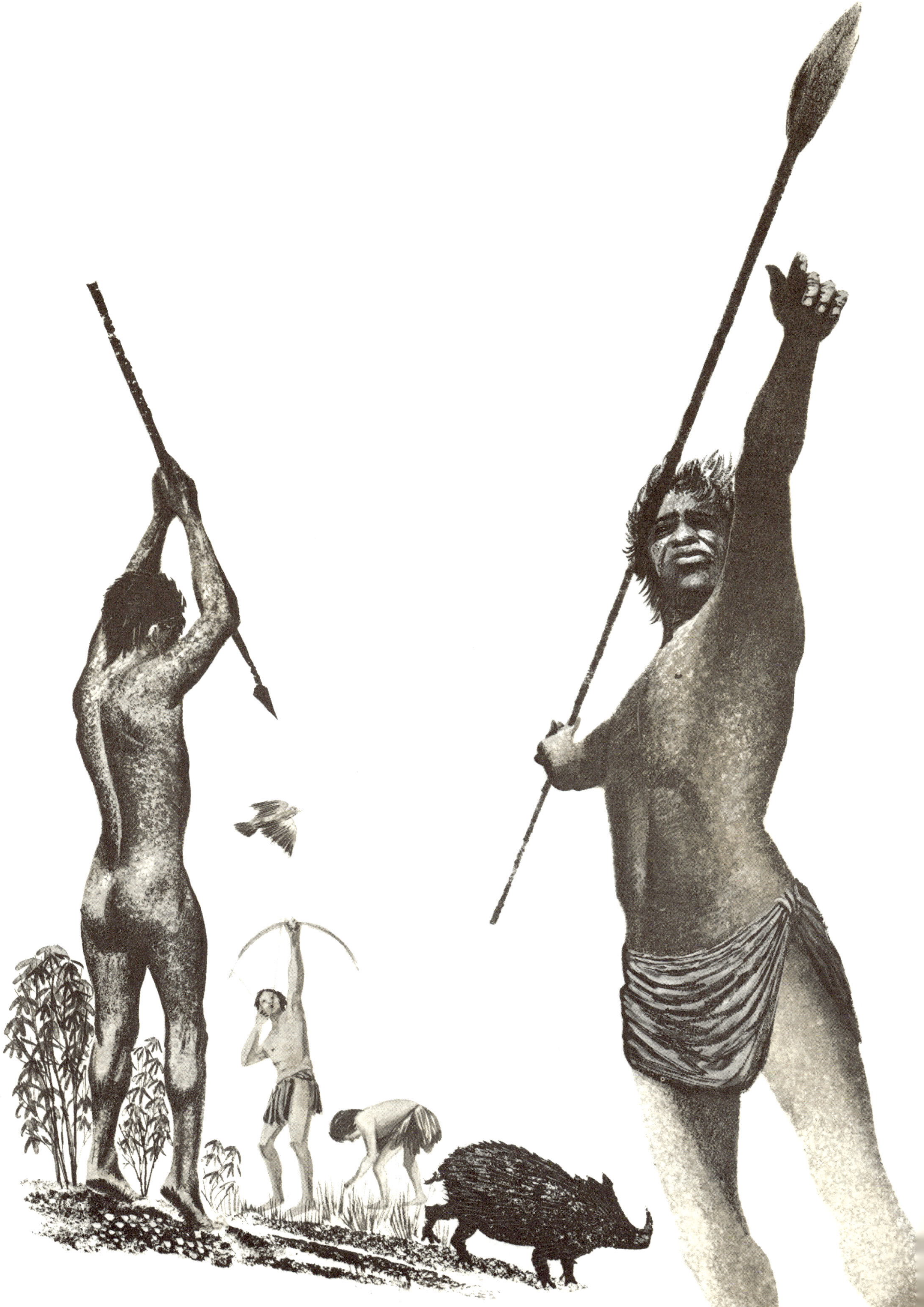

THE FIRST FARMER

One of the biggest changes in the history of man took place about 9,000 years ago. Up to that time, in order to stay alive man had to move about constantly in search of food. He lived in an often desperate hand-to-mouth fashion—hunting, fishing, collecting berries and grain, whatever was at hand. The women were the gatherers of food; the men were the hunters and fishermen. People could not control what they had to eat nor decide when they would have it. Like animals and birds, they were at the mercy of nature.

Then man took his first giant step toward civilization. He found he could grow food. No one knows just how this discovery was made. Perhaps someone noticed plants growing where seed had accidentally dropped. Perhaps someone had stored wild wheat and barley in rock-shelters and later found some of them had taken root. The important thing was that now for the first time and in small ways nature could be controlled. People were more independent; they could stay in one place and did not have to give all their time and thought to the search for food. Man had become a farmer. But men as individuals were not the farmers. Women had always been the seed gatherers. Now they sowed the seeds, tended the plants, harvested the crops. In fact, women were the first farmers.

Clay impressions found at Jarmo

ARABIAN SEA

Where Man Farmed

The discovery was made in different places at different times. Probably the first farmers planted their crops in the foothills of the Zagros Mountains along the border between present-day Iran and Iraq. This part of the world seems to have been particularly suited to the beginnings of farming.

In earlier ages it had been ice-covered. Then over a long period of time the glaciers melted away to the north. After the glaciers receded, the climate became temperate with rainfall enough to produce crops. There is evidence that wheat and barley grew wild here. Man was already using these wild grains for food. A great variety of fruit—almonds, peaches, apples, pears, figs, plums, dates, olives, and apricots—all grew in this fertile region. Many of the animals that were later to be domesticated roamed wild here.

The conditions were right. Here man could thrive and settle down.

Later, and perhaps independently, other early farming communities developed in Africa, China, and Central and South America.

WHEATS

Wild   Cultivated

BARLEYS

Wild   Cultivated

EARLY FOOD CROPS

Once he had discovered the principle of farming, man planted from seed what he found growing naturally. Wheat and barley were among the first cultivated crops. The best crops were those which ripened quickly over a period of a few months—oats, rye, and beans. These fortunately also kept well in storage, insuring a food supply over a period of time. With fruit and nuts, which grew abundantly, the early farmer had a variety of food.

While men hunted, fished, caught birds, women collected nuts, berries, edible roots, seeds and honey.

Seed saved from the last harvest was scattered over prepared ground to take root and grow.

How Man Farmed

Man now set about deliberately to farm. Solid ground must be broken up before seeds could sprout and take root. In earlier times food gatherers had used the digging stick to pry out edible roots. The farmer found he could use this same tool to break up soil. Over the ground he had prepared, he scattered wild seed collected and kept from the year before.

People who lived along streams had less of a problem. Ground was prepared for them naturally by spring floods which deposited layers of rich loose soil. Sheep, pigs, small hoofed animals, were sometimes used to trample the rich alluvial soil and press down the seeds.

For seeds to grow well farmers had to learn to plant them far enough apart to dig out weeds and to guard the growing plants from bird and animal marauders.

When the grain was ready to harvest, the farmer cut the plants and dried them.

Using stone-weighted digging stick to break ground for planting

The next step was to separate the edible seed from the straw. The simplest method was to spread the dried grain on hard ground and beat it with sticks. This loosened the seed from the chaff. Also animals were beginning to do the work of men. Driven around over the grain, they loosened the seed just as man did by beating it.

The farmers tossed the seed and chaff into the air to separate them. Grain seeds, heavier than the chaff around them, dropped to the ground. The chaff was blown away. As time went on, threshing floors were set aside as special places for separating the grain. The grain was used for food, the chaff for many purposes, and the straw for fodder.

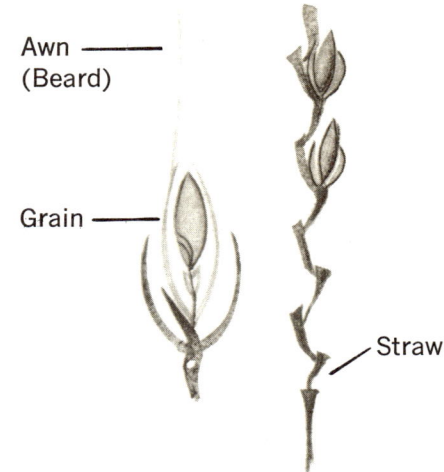

Head of wheat
Chaff is everything but the grain itself.

Man's Uses of Plants

Now that man could farm, new ways of life opened up to him. It was possible to have a permanent home instead of temporary shelters. As he built and as he made clothing and household equipment, he could now make use of the plants around him. Wheat and barley stalks could be used for bedding and for thatching roofs. Flax and hemp provided fibers that could be woven. Some plants were effective as medicine; some were used for paint, for building material, for baskets and matting. From squashes, gourds and pumpkins, dishes, scoops, cups and ladles could be made.

Now man could acquire possessions. He was taking further steps toward civilization.

Foods were pounded and ground with a grinding stone on a quern. Early farming people ground grain, then baked it, making flat unleavened loaves — or boiled it making porridge.

The wolf may have been dog's early ancestor. From as far away as the Baltic Sea region and ten thousand years ago, evidence of domestic dog has been found. A wild species of horse was native to the cold grasslands of central Asia and was first domesticated for its milk and meat.

From somewhere in southeast Asia the pig, a descendant of the wild boar, moved into forest-covered Europe and became a most important source of food. Knives and ornaments were fashioned from pigs' tusks.

The ancestor of today's cattle was the aurochs, an enormous, fierce animal with long spreading horns.

Man Domesticates Animals

Early man not only brought plants under his control, he also gradually tamed animals. There is evidence that before man learned to farm, dogs in search of food and warmth had come to live with people. Indeed, goats and sheep seem to have been among the first animals to have been herded in the days when people were still wandering from place to place.

These same early people had learned to use goatskins as containers. As they traveled, they found that milk carried in these containers bounced and jiggled and soon turned into butter.

Sheep and goats supplied man with meat, milk, butter and cheese. Goatskin was used for tent shelters, clothing, containers—the wool for woven materials.

Those early people had found that animals were a source not only of food but of many other products. Straps, thongs, shields, were made from cattle hides. Spoons, cups, containers, were fashioned from horns.

No one knows for certain how Neolithic man was able to tame large herds of animals. While man still had to wander in search of food, he already took with him herds of sheep and goats. Hunters may sometimes have brought home baby goats and lambs and raised them as pets. Bones of sheep, cattle, pigs, and goats have all been found in the earliest settled communities.

Studies of animal bones show that over a period of time there were changes in structure and appearance. Coats changed, too. Man was learning to breed his domesticated beasts in ways which made them more useful to him. They put on more flesh. Their bones and teeth grew smaller and their tails shorter. The thin coat of the early wild sheep through breeding grew woollier and more abundant.

Ox skull, Jarmo

Aurochs horns found at shrine, Çatal Hüyük

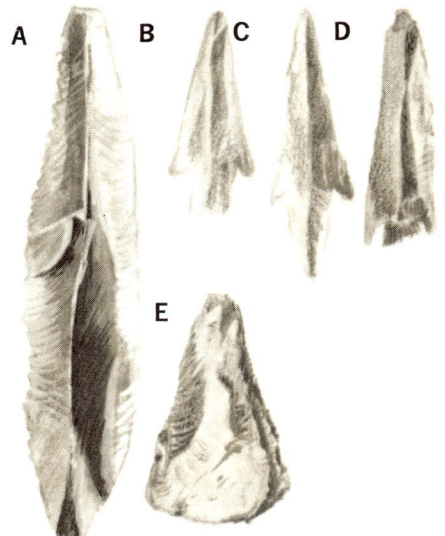

FLINT IMPLEMENTS
A Flint sickle probably hafted as a knife, Jericho
B, C, D Flint implements of the pre-pottery Neolithic period, Jericho
E Flaked-stone hoe blade, Hassuna

Tools

The earliest hunters of the Old Stone, Paleolithic, Age used stone to make tools: knives, axes, scrapers, borers, tools with which to cut, to strike and to pierce. Where bone or wood could be found, they were used to make fishhooks, polishers and awls.

To chip off a piece of stone suitable for a tool, man used a technique called percussion flaking. Sometimes he smashed two stones together, often hurling a large chunk against an outcrop of rock to shatter it. Sometimes he struck one stone with another heavy hammerstone. This crude hit-or-miss method produced only a few usable fragments.

Later he found he could get small chips by pressing hard against a stone with a bone or antler prong. This was called pressure flaking. Still later he made blade tools by splitting slender prisms of stone from a stone core.

Stone adze blade in antler sleeve, set in wooden handle, lake dwellings, Switzerland

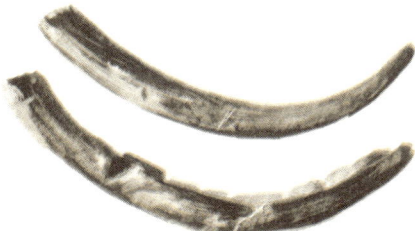

Deer antler sickles, Hacilar, 8000 B.C., one still fitted with chert-stone blades

Bone implements, Çatal Hüyük:
A Spatula, **B, C** Spoons, **D** Fork

Making a hand ax with a bone hammer, here the bone of a horse.

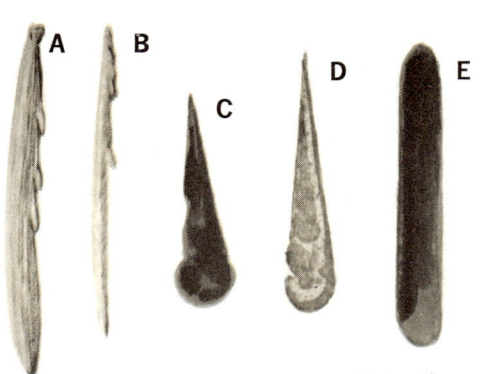

A, B Very early harpoons, Palestine
C, D Two bone awls, Dalma Tepe
E Whetstone, Dalma Tepe

A Ground stone quern, Jarmo
B Stone bowl, pre-pottery Neolithic, Jericho
C Ground-stone vessel, Lower Egypt

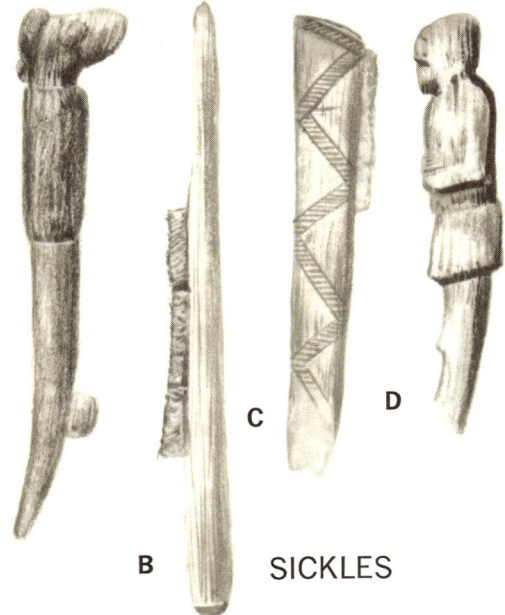

SICKLES

A Very early bone, Mt. Carmel, Palestine
B Bone, lower Egypt
C Bone, northern Iran
D Bone handle, northern Iran

Pre-pottery Neolithic quern
Jericho

Ground-stone quern
Northern Iran

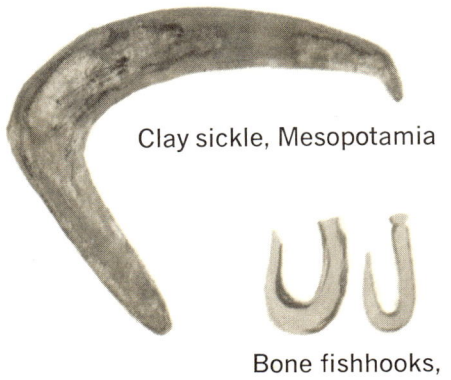

Clay sickle, Mesopotamia

Bone fishhooks, probably oldest known
Palestine

Hammerstone

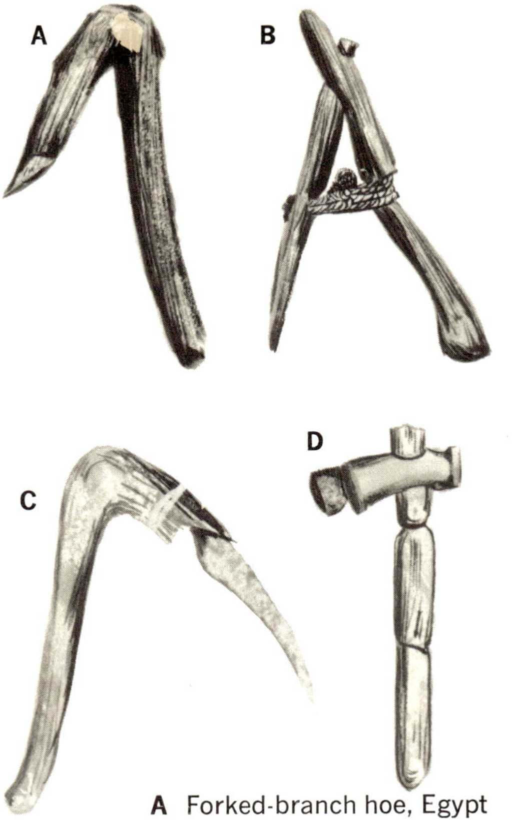

A Forked-branch hoe, Egypt
B More developed hafted wooden blade, Egypt
C African hoe
D Ax blade in a deer-antler socket, France

Polished stone axes, France

As man learned to control nature he learned also new ways of shaping stone. By grinding and polishing he could make better-shaped tools with a finer cutting edge which could be sharpened rather than discarded when blunted by use. He produced this edge by rubbing the stone tool with abrasive sand, with sandstone, or with harder stone.

Man used these blades in a variety of positions and each position gave him a new tool. Gradually the hoe had replaced the digging stick; now the hoe became an adz. Man used his tools with new skill. He could clear ground for new farmland, shape timbers for use in building and even make furniture.

Over a long period of trial and error, man found that flint was one of the best stones for his purposes. The best flint was that dug up from the ground. Flint found lying on the ground exposed to air was more brittle and did not produce the best tools. Early man had sometimes to travel long distances for flint. Digging into chalk cliffs, he mined for it. There is evidence of large-scale tunnel mining and rather elaborate mines with sunken shafts and connecting galleries.

One of the centers of flint quarrying was at Grand Passigny in France. The flint there had a distinctive color like amber or beeswax. There is evidence that it was carried great distances. Flint of this color has been found in Italy, Switzerland, and Belgium.

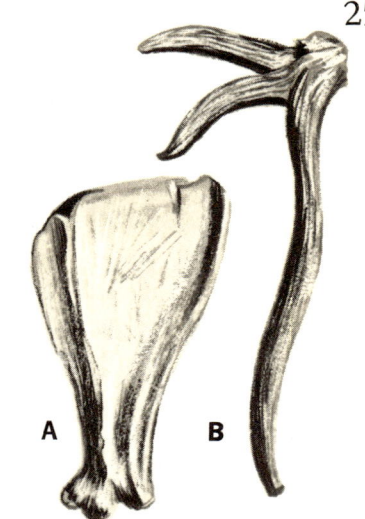

Mining tools, England
A Shovel, shoulder blade of an ox Sussex, England
B Antler rake, Sussex

Vertical cross section of flint mines showing open workings and later shafts and galleries, Spiennes, Belgium

Deer-antler pick England

Horizontal cross section of pit shaft with radiating galleries, England

Cave dwelling

Homes

From earliest times people sheltered in caves under rock ledges and in crude makeshift temporary shelters. Now that they could grow plants for food, people could settle on the soil and build more permanent homes.

In different parts of the world people built different kinds of houses using what material was at hand. Their homes were adapted to the climate and conditions they lived with.

The windbreak, one of the earliest types of shelter, was made of trees or branches set into the ground and covered with brush, grass, or skins. Later houses were woven of upright sticks twined together with flexible branches—wattle—then plastered with mud—daub. The roof was thatched with straw.

Mud brick

Open windbreak

Domed stone round houses

Wattle and daub

Lake dwellings, Switzerland

Long house, Rhineland

European lake dwellers built wooden houses on piles driven into the beds of lakes or marshes. At one site in Switzerland remains of 50,000 piles have been uncovered.

In the great river valleys mud was roughly patted to make the walls of a house. Bricks could also be made of sun-dried mud.

Where mud was plentiful, as at Jarmo, Jericho and along the lower Nile, houses were made of "tauf"— sun-dried clay mixed with straw and grass to prevent cracking. Animal droppings were used as a binder to hold all the elements together. Where there were enough large stones men piled them up for foundations or walls. Later, in places where trees were plentiful, people built houses of logs or of poles.

Fragments of daub wall showing impressions of wattle
Italy

Thumb-impressed brick
Jericho

Villages grew up close to the farmers' fields. With a larger, more certain food supply the population increased, too. There were more people to do more things beyond the demands of mere survival.

Life was less hazardous. More infants survived and grew up to share the work of the community. Boys shepherded herds and flocks. With the women, girls planted, cultivated and harvested.

Toward the end of the Neolithic time there were houses with several good-sized rooms, separate cooking areas, hearths and built-in ovens. Some rooms contained built-in beds, closets and plaster cupboards with shelves. There were spacious courtyards, storage rooms, sheds, and pens for livestock.

Houses, storerooms, courtyards and shrines, about 6500 B.C., Çatal Hüyük. Entrances only through roofs.

Post and lintel—a horizontal length, wood or stone, placed across two upright supports, or posts—a way of supporting weight above an open space developed by Neolithic builders. This structural form was basic to the growth of architecture.

The tower of Jericho, showing the highly developed technical ability of early Neolithic builders.

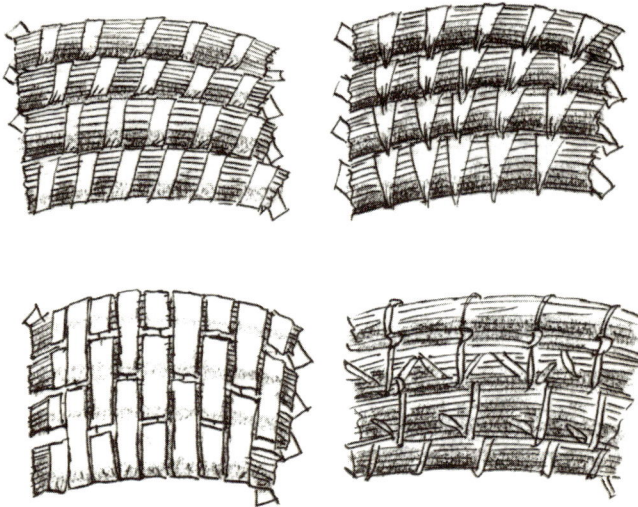
Different ways of coiling

Basketry

Man has always needed containers in which to carry, store and preserve things. Although hunters and food gatherers used a variety of materials to make containers—skins, netting, leaves, bark, fiber, clay, stone, shells, gourds—archaeologists have proved that baskets were among the earliest and most often used. No one yet knows how the discovery was made that certain grasses or reeds could be twisted and woven. Excavations do show that in the ancient world there were two major ways of making baskets: by coiling around and around and sewing the coils together, and by weaving.

Ways of making baskets have changed little through the ages. From the earliest days strands of fibers were woven together by hand to make a basket, a mat, or a material. Man used whatever plant fiber grew close to hand. Natural fibers are perishable. Only scattered fragments of basketry can sometimes be found today. These survive mostly in dry desert regions.

Braided basketry, sewn

Ways of starting to coil

Boat-shaped basket, Lower Egypt
Length about 16 inches, 4500 B.C

Coffin of woven rushes, Egypt
Length about 47 inches

WAYS OF MATTING

Rush matting
Egypt

Bed matting
Egypt

Single reeds

Bundles of grasses

Reeds on cord

Spindle and whorls

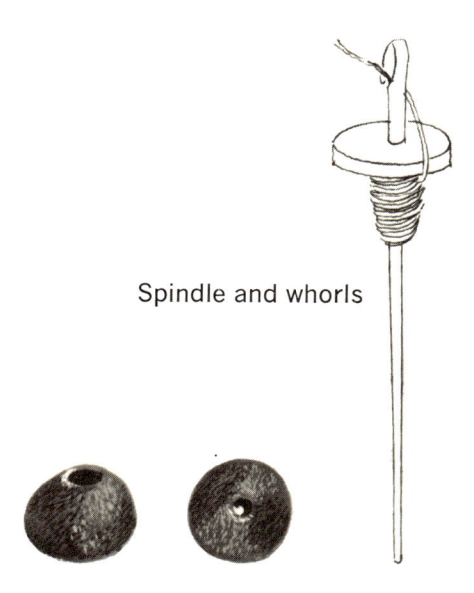

A spindle whorl was a disc with a hole at the center which helped control the twirling spindle.

Spinning and Weaving

Long before farming began, the basic techniques of weaving had been discovered. Coarse thread was made by twisting together human or animal hair—of wild mountain sheep, goats, dogs and others. The spindle—for spinning thread—and a weaving frame—for weaving thread into cloth—were used.

The crops the farmer now grew provided new material for thread—flax, cotton, hemp.

Textile adhering to bone
Çatal Hüyük

Fine shawl-like material
Çatal Hüyük

Spinning wool
The spindle is weighted with a whorl.

Working with this variety of materials, weavers produced more and experimented also with methods that would make weaving easier. One simple but important improvement was placing the weaving frame in a horizontal rather than a vertical position.

World's oldest known textile
About 6500 B.C., Çatal Hüyük

CLOTHING

The hunters of the cave period depended upon animal skins for clothing. But raw hides soon stiffened and hardened. They had to be cured—softened and preserved. Skins were softened by alternately wetting and beating them.

Then man discovered that skins could be softened better by rubbing with animal fats, brain and marrow, urine or dung.

Later he learned to tan skin by soaking it in water with the bark of oak or willow tree. The water and bark produced the tannic acid which cured the hides. This process turned raw hide into leather.

Skins were still used for clothing but now with new materials and methods of weaving there was cloth, too.

Material for weaving cloth came from the wool and hair of animals and from plants—hemp, flax and cotton. The material available varied from place to place and from climate to climate.

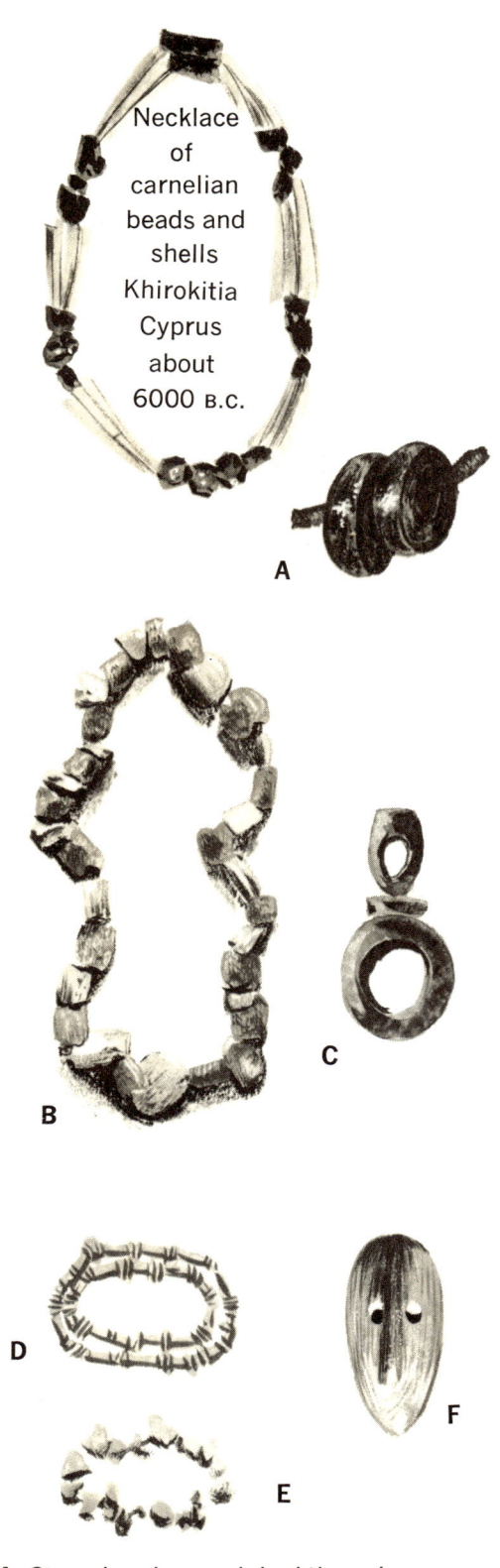

Central Asiatic people learned to make felt. This material is a combination of hair and wool. It was made by first combing out the raw materials and placing them in layers on a mat. The materials were sprinkled with water. Then the mat was rolled up, beaten or rolled back and forth until the hairs and wool fibers inside became thoroughly matted. Patting, stretching and repeated rolling produced felt. This light, warm, durable material was used for garments, headgear, boots, and for tent covers and rugs as well.

The exact kinds of garments worn by men and women are very hard to determine from archaeological evidence. What we do know can be learned from wall paintings, figurines, pottery fragments and on rare occasions the very impressions or tiny fragments of material which have survived to this day.

The commonest articles of clothing were loincloths for the men and for both men and women aprons, simple skirts and shifts, sometimes belted.

In addition to garments necessary for comfort and protection from the weather, man ornamented himself for various reasons. Hairpins, ribbons, necklaces, bracelets, anklets, rings, combs, pendants, amulets: these articles were worn for pleasure but sometimes as signs of importance and some of them as symbols with magical significance. These were made of wood, bone, stone, shell, clay, seeds.

Gold, turquoise, copper, ivory and amber were rare and probably used only where they could be found by chance or acquired by trade.

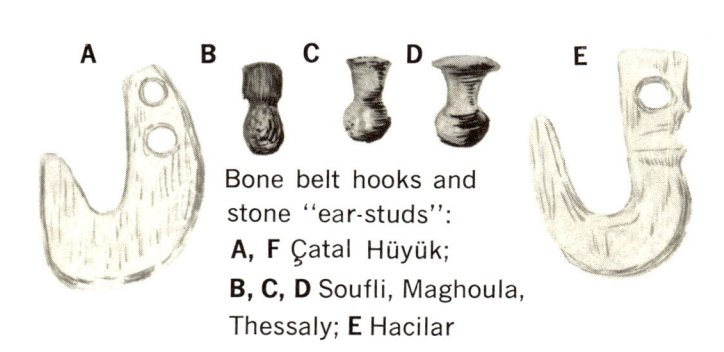

A Stone beads on original thread Çatal Hüyük, about 6500 B.C. Shown nine times actual size
B Necklace of blue and amber-colored beads, Çatal Hüyük, about 6000 B.C.
C Bone belt hook, Çatal Hüyük, 6000 B.C.
Bracelets, **D** beads, **E** polished stone, Çatal Hüyük, 6000 B.C.
F Bone face-bead, Jericho, before 5000 B.C.

Bone belt hooks and stone "ear-studs":
A, F Çatal Hüyük;
B, C, D Soufli, Maghoula, Thessaly; **E** Hacilar

Tumpline, a band across the forehead to support weight, carried on the shoulders and back.

Transportation

The farmer could inherit from the hunter some of his basic methods of transportation because a hunter had to travel extensively in search of food.

Man himself was the carrier and he developed various devices for carrying heavy loads—first a combination of poles and yokes. Then he dragged loads on animal skins. Later flat wood pieces replaced skins and made the sledge.

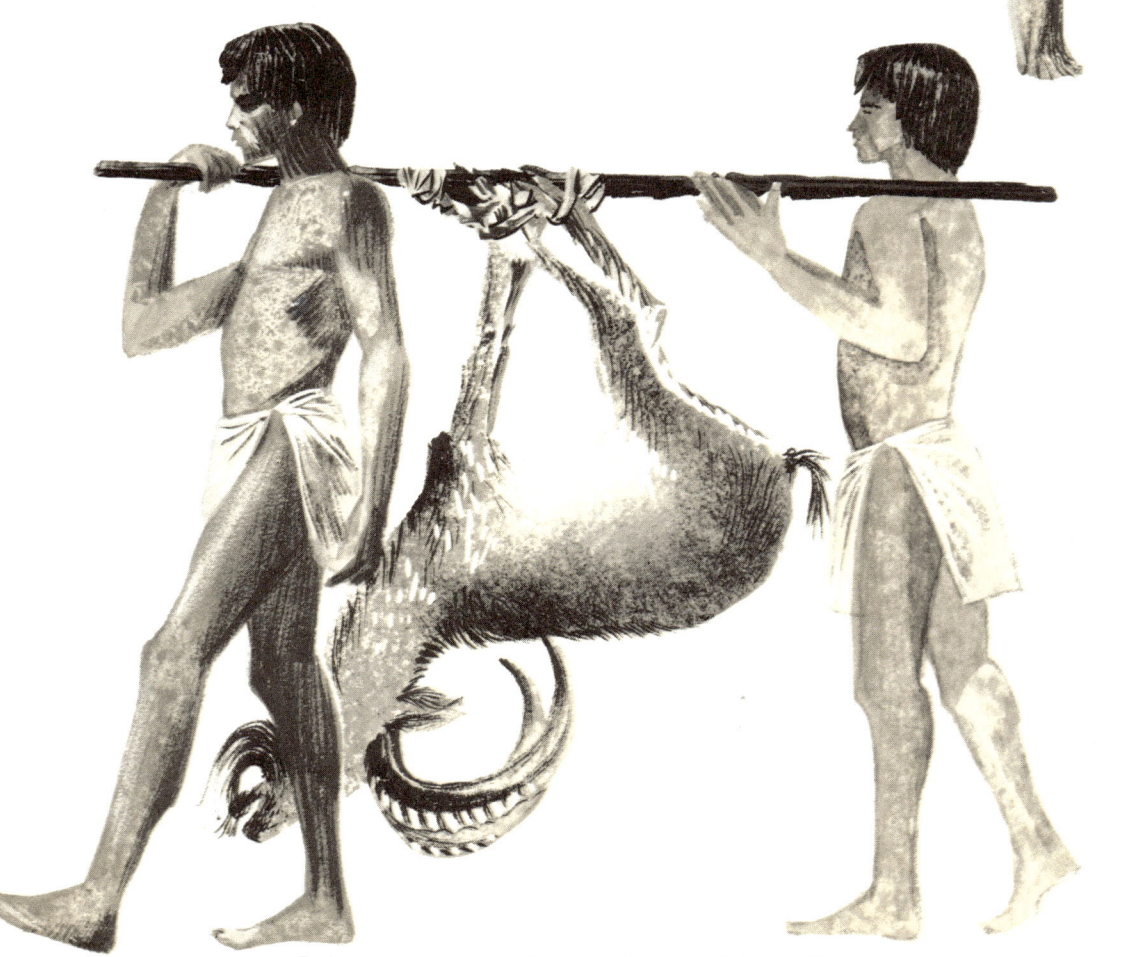

Pole support carried across the shoulders of two

V-shaped sledge cut from tree fork

Transporting water in animal skin

People who lived near water discovered very early that they could use logs, gourds, or inflated skins as floats. They made rafts of bundles of dry reeds fastened together. Sometimes large basket tubs were covered with pitch and floated in the water. Dugout canoes were common. These were single logs hollowed out. Sometimes birch bark was sewn over wooden frames and the joints were filled with plant gums to make the vessel watertight. In shallow waters men used poles to propel their craft, but in deeper water, such as the Mediterranean, they used paddles or oars. Some of these early vessels contrived by men were capable of far travel. Neolithic culture was brought from the mainlands to the islands of the Mediterranean, and by the end of the period sea travel reached out to much more distant places.

Man began exchanging some of his goods for raw materials and things that were made or grown in other places. As people visited and traded, they learned from one another. Man's interests broadened; his ways changed and improved.

Dugout

Boat made of reeds

A quffa, bitumen-covered circular boat
used today on the Tigris

Clay boat, Eridu, Iraq
10 inches long, perhaps a toy

Coracle
a basketlike frame
covered with skin

Raft

Early Neolithic pottery shards
Çatal Hüyük

Beating clay on inverted old pot

Pottery

A pottery fragment, or shard, is one exciting clue archaeologists use to reconstruct the lives of early peoples. Lumps of clay could be worked into various useful shapes. Even in his hunting days man had found that certain clays dried and hardened in the sun.

Clay for making pots had to be prepared in much the same way as for making bricks. Straw was added to prevent cracking and dung to give it strength. To remove moisture the mixture was pounded and beaten. The clay was then ready to be shaped in various ways.

Basket-handled jar of brown burnished ware, Çatal Hüyük about 8 inches high, 5900 B.C.

Cooking pot with four lug handles Çatal Hüyük, about 10¼ inches 5900 B.C.

Coiling clay

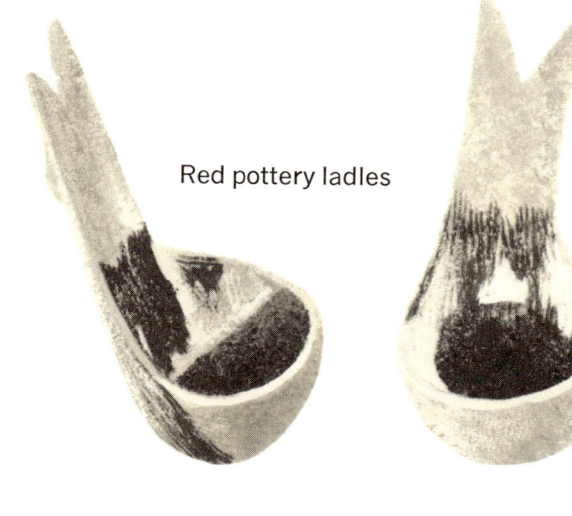

Red pottery ladles

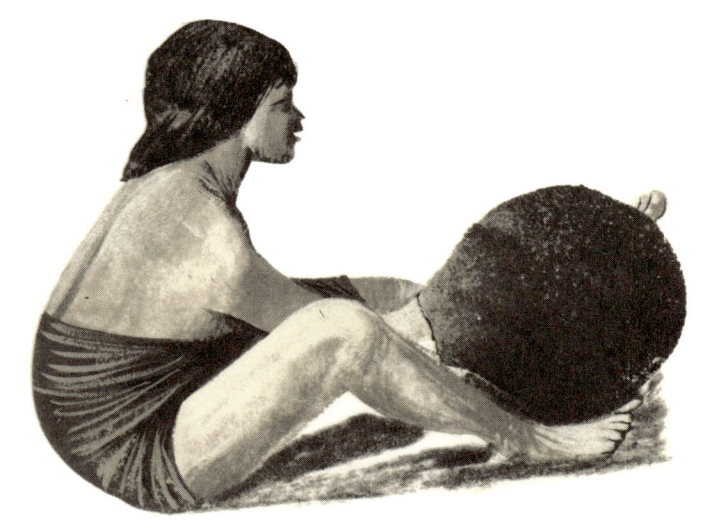

Forming a pot by beating inside with a wooden pestle

Neolithic clay pots imitating leather bags in shape and decoration. Michelsberg, South Germany

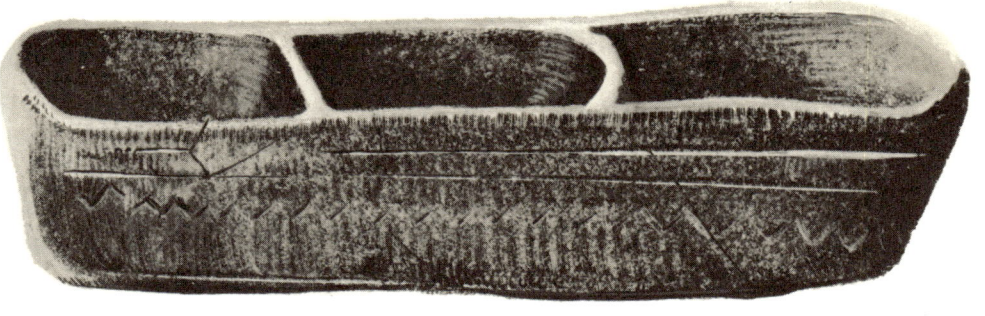

Neolithic trough, Knossos, Crete

Neolithic jar Northwest Europe

Man took a big step forward when he learned to fire clay instead of just leaving it to bake in the sun.

He discovered that baking clay a long time in a hot fire hardened it so that it was leakproof. Fired clay pots could be used for cooking, storing, and transporting liquids.

For baking clay he dug trenches and piled the pots he had shaped into them. He covered them over with a thick bed of branches, straw, wood, and dung and set fire to it. The smoldering mass baked the clay at a high temperature. The trench and surface layers of straw and wood kept the heat in. The fire was allowed to die down before the pots were lifted out of the trench. A red-hot vessel taken directly out of intense heat was likely to crack.

The advance in the use of clay for both containers and building materials was one of the most important achievements of Neolithic man.

Plain ware pottery
Straw tempered, poorly fired
Hajji Firuz Tepe, Persia

Stack of pots being fired, present-day Nigeria

Animal vase in the
form of a boar, Hacilar

Inverted cup in the form of
a woman's head, Hacilar

Animal vase
A ritual vessel in form of a deer
Hacilar

Painted ware
Rachmani, Thessaly

Painted ware
Dimini, Thessaly

43

44

Earliest painted pottery
Red on cream, Hacilar

Shards of earliest painted pottery
Tepe Guran

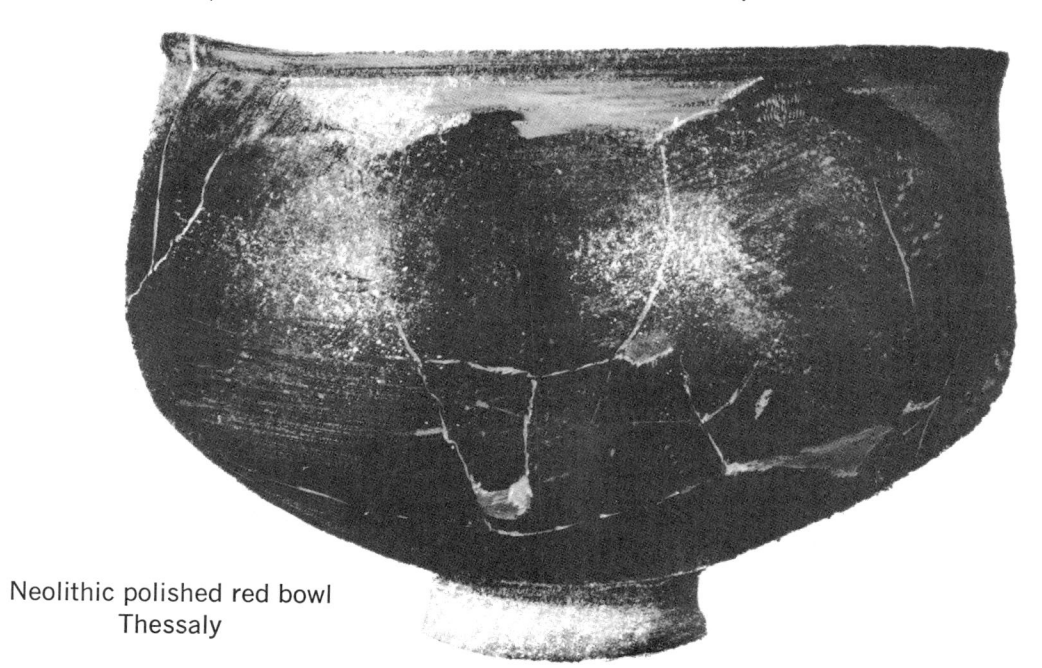

Neolithic polished red bowl
Thessaly

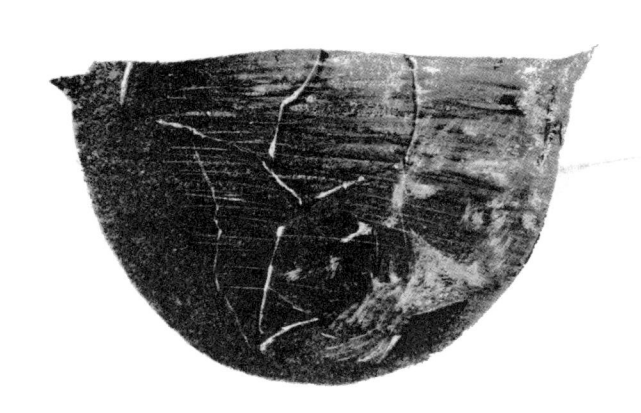

Early Neolithic dark burnished pots decorated with cardium shell combing, Byblos

Painted pottery
Hassuna

Painted pottery
Mersin

Art

The hunter left enduring pictorial records. His paintings on cave walls have survived through ages of time. These caves were opened and the paintings have been known for at least 100 years.

Evidence of the arts of early farmers has until recently been limited to fragmentary remains of basketry, weaving and pottery. The very process of weaving, whether basketry or cloth, naturally produced geometric patterns. It is possible that these carried over to the designs on early clay pots, tools, to anything he might make.

Then farmers went on to illustrate their observations, feelings and reactions toward what they saw about them in their daily lives—the sun, rain, lightning, plants, animals and people—in painting, in pottery, in all their artifacts. Clay figurines, tools and other objects were designed in new ways which were not only decorative but had meaning. Some may have been amulets or charms used for good fortune or for plant, animal, and human fertility.

Cave painting, charging bison
Pre-Neolithic, Altamira

Cup with animal handle
Çatal Hüyük

Embracing couple
Earliest stone sculpture
8500 B.C., Ain Sakhri

End of bone spatula
carved bull's head
Hacilar

Early stone head
Eynan, Palestine

Early stone head
Eynan, Palestine

Stone figures from Çatal Hüyük, 6500 B.C.: hooded, cloaked figure of black limestone; kneeling goddess, white marble; young god, white marble.

What was known of the art work of early farmers has completely changed because of the discoveries made in excavations at Çatal Hüyük.

Girl with braided hair resting in position common among primitive people, Hacilar, Late Neolithic, 5500 B.C.

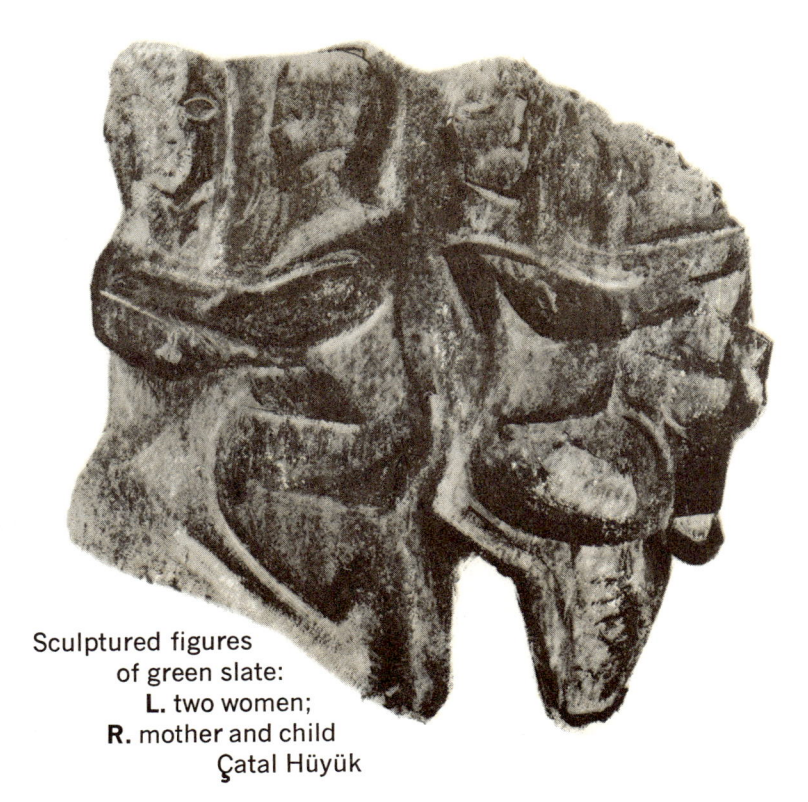

Sculptured figures of green slate:
L. two women;
R. mother and child
Çatal Hüyük

Murals from a shrine room, Çatal Hüyük — 6000 to 7000 B.C. — recently uncovered by James Mellaart, lecturer in Anatolian Archaeology at the Institute of Archaeology, University of London.

Until the discovery of these murals no evidence had been found that men of the New Stone Age painted on such a large scale, with such skill and dramatic concept, as had the men of the Old Stone Age, especially at Lascaux and Altamira, more than 20,000 years ago.

Above is a reconstruction of the enormous wall painting of an aurochs bull. The bull figure is six feet long. Around the awesome animal are fragments of figures which would seem to have been hunters or worshippers.

To the right is a reconstruction of the deer hunt mural from the same shrine room.

49

RELIGION

Primitive man was very much a part of nature. For him the world was filled and bursting with life. Every man or woman, plant or animal was just as much an individual as he was himself. Even an object such as a stone or something that happened in nature like a flood or a thunderstorm was alive. He did not give the animal or stone qualities like his own. He believed they already had them.

As a hunter, man's first concern was with the supply of animals and with his own ability to trap and kill them. Success in the kill was dependent upon rituals and magical figures which he believed could protect him against harm and make things happen favorably for him.

Leopard relief-carved stone wall in shrine

Goddess with leopard, stone sculpture, height 4⅜ inches, Çatal Hüyük, 6500 B.C. Anatolia

All of his world, he believed, was at the mercy of mysterious forces. He created a multitude of gods or more often goddesses who personified them. Because these powers were unknown and mysterious, man feared them. He felt the need to do something to get them on his side.

The fertility of the farmer's fields and herds, the well-being of his family, the fertility of his wife, all depended on his ability to win over these mysterious forces.

In the farmer's personal world the thunderbolt and rainstorm were male. His fields, the earth which gave birth to plants, were female. In the animal world the bull was a symbol of strength and power, of male virility.

Remains of north wall of a shrine
Lower part of bull cut into stone
Çatal Hüyük

Reconstruction of north wall of shrine showing black bull and hand patterns

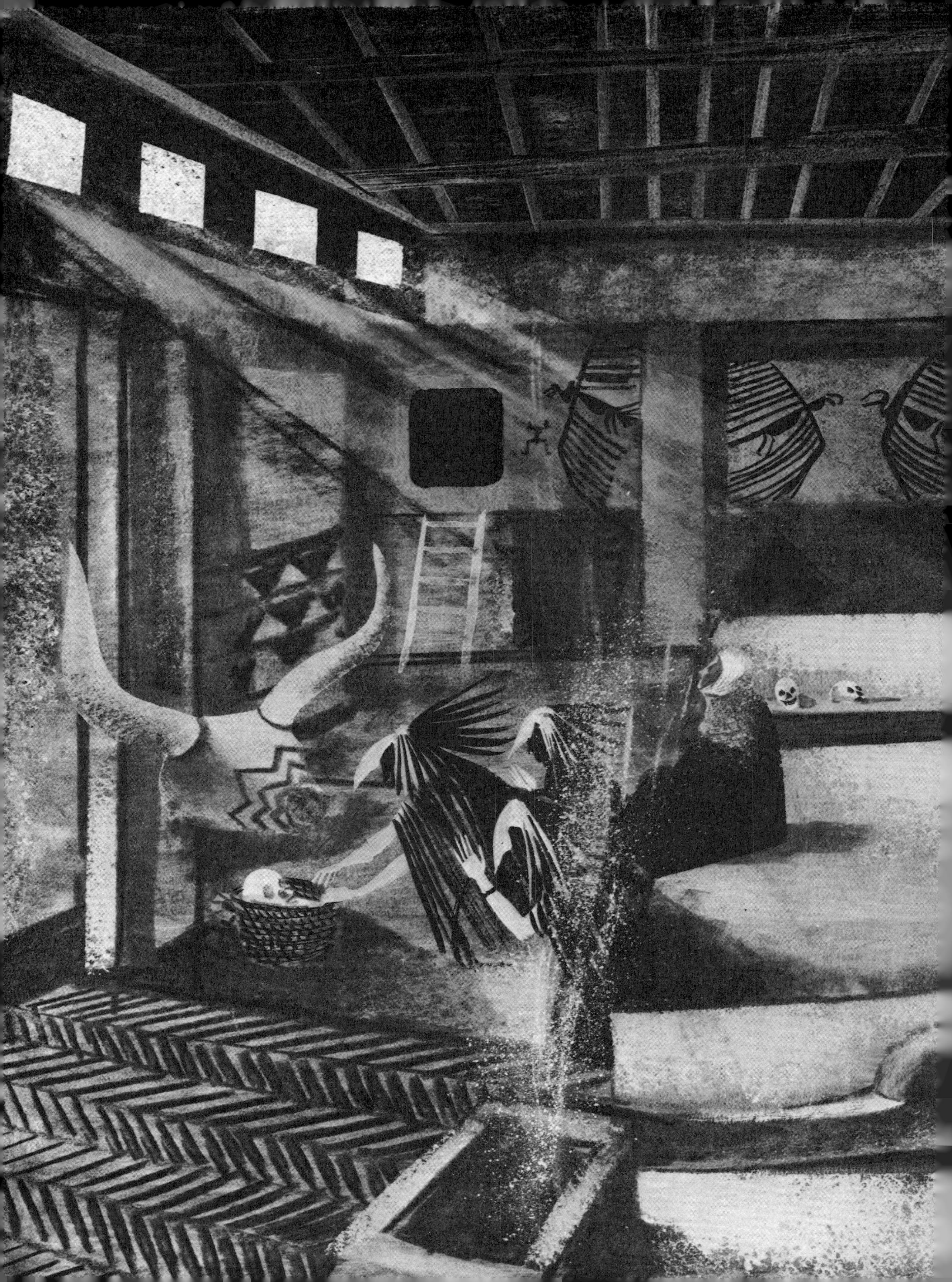

Just as the hunter had painted on the walls of his cave figures of animals which became a part of rituals to bring him success in the hunt, the farmer now fashioned figures which he believed would influence mysterious forces on his behalf. For the same reasons, he made and wore clay or stone amulets of many sorts for good fortune and protection against disaster and death.

Human bones have been found with objects of everyday life buried with them. The farmer's burial customs may have been intended to assist a spirit which lived on beyond death or he may have felt that the spirit like other mysterious forces might work him either good or harm.

Neolithic skull
with features modeled in plaster
Shells used for eyes, Jericho

Reconstruction of a funeral ceremony in a shrine room based on actual discoveries, Çatal Hüyük, 6150 B.C.

after Alan Sorrell

CONCLUSION

What Neolithic man achieved is even more a miracle than it might seem. It could have happened only at a time and a place where climate and land were suitable for growing plants and domesticating animals in abundance. Five times at most this combination occurred in prehistoric times: in the foothills of the Zagros Mountains, in the Nile Valley and in the Indus basin, all three in the region we have been exploring. Once it occurred in Central America and once perhaps in Peru. It is important here to understand that the roots of modern, Western culture, of which our present-day civilization is a part, did not come from the Neolithic Revolution in the Western Hemisphere. Civilization in America, coming to us by way of Europe, had its roots in those first farming communities around the eastern end of the Mediterranean.

The miracle was what man did with that special moment—for although the period was thousands of years long, in the evolution of the earth and of man it was a very small span of time. What Neolithic man did made it possible for more people to live together securely with greater abundance in a more permanent home.

Out of the settled communities of those first farmers, those domesticators of animals, came a much closer association of human beings and a need to develop new patterns of community relationships. Out of those communities grew the cities, rich breeding grounds of man's talents as craftsman, trader, artist, thinker and builder of the worlds to come.

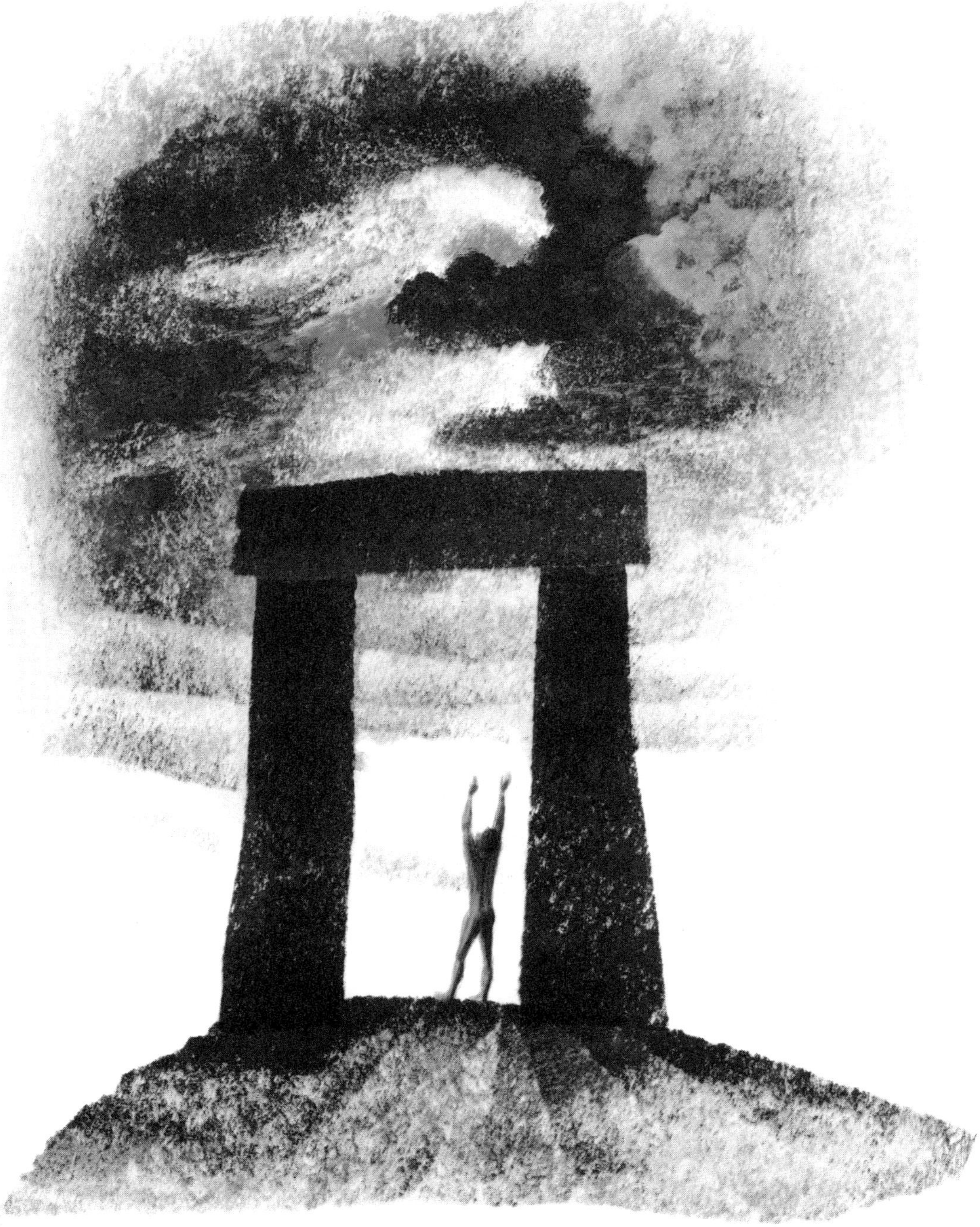

N
ITALY
THESSALY
BLACK SEA
CASPIAN SEA
ANATOLIA
Sesklo
Çatal Hüyük
SYRIA
MESOPOTAMIA
IRAN
Jarmo
ZAGROS MTS.
MEDITERRANEAN SEA
Khirokitia
IRAQ
Jericho
NILE R.
EGYPT
ARABIA

IMPORTANT SELECTED NEOLITHIC SITES
Descriptive Material Prepared by Edward Ochsenschlager

JARMO
Jarmo in Iraq was excavated by Robert Braidwood. It appears to have been a small village of about twenty-five houses, each with several rooms. Many unbaked figurines of sheep, dogs and sitting women of "mother-goddess" type were found, as well as stone vessels and tools, pottery, flat stone palettes for grinding ochre, and small objects such as bone pendants, awls, spoons, pins. The lowest radiocarbon date for Jarmo is 9627 ± 309 B.C.

JERICHO
Kathleen Kenyon's excavations in Israel have added a great deal to our knowledge of Neolithic Jericho. Jericho is the earliest fortified town that archaeologists have discovered. Huge walls, five feet wide at the top, reached a height of over sixteen feet. A large tower, twenty-seven feet thick and still standing twenty-one feet high, served as a lookout post where guards could scan the surrounding countryside and warn the inhabitants of approaching danger. One sample from the pre-pottery Neolithic A, the period during which the fortifications were built, is dated by radiocarbon to 7114 ± 165 B.C. The earlier Natufian period has been dated by radiocarbon to 8196 ± 247 B.C.

ÇATAL HÜYÜK
Çatal Hüyük in Turkey, excavated by James Mellaart, was a Neolithic town more than twice the size of late pre-pottery Jericho. Excavations on this site have revealed a vast amount of detailed information about all phases of life in Neolithic times, more than any other excavation in Anatolia. Apart from standard finds of bone, stone, and pottery, Çatal Hüyük has yielded paintings, painted reliefs and modeled animal heads used as wall decorations, which enrich our knowledge of the art and religion of this period. Houses were built here with timber reinforcements and decorations. Radiocarbon dates range from 6385 ± 101 B.C. in level X to 5797 ± 79 B.C. in level II.

KHIROKITIA
P. Dikaios, the excavator of Khirokitia on Cyprus, has uncovered forty-eight domed huts with square limestone pillars supporting a partial second floor and containing niches or cupboards for storage. It is thought that nearly a thousand of these huts existed in Neolithic times occupied by several thousand inhabitants. People of this very large village were familiar with spinning and weaving, raised sheep, goats, and probably pigs, practiced agriculture and, although they made a few early experiments with pottery, used polished stoneware for containers. Samples from Khirokitia have been dated by radiocarbon as early as 5650 ± 150 B.C.

SESKLO
D. R. Theochares has recovered an abundance of information about the Neolithic period in Greece through his excavations at Sesklo. Here pre-pottery levels are succeeded suddenly by levels where pottery of expert workmanship was made. In one middle Sesklo period in Thessaly archaeologists have discovered the earliest known *megaron* houses, with a small porch leading into a great central hall. This type of building survives through the Bronze Age, appears in early Greek architecture and constitutes an important element in the classical temples of the Greeks. The Neolithic period at Sesklo extended from about 6000 to 3000 B.C.

GLOSSARY AND PRONUNCIATION GUIDE

AIN SAKHRI (añ sákri) — A site in Palestine.

ALTAMIRA (äl′ tä mē rä) — Site of famous Paleolithic cave paintings of the Magdelanian period in northern Spain.

ANATOLIA (an ə tō′ li ə) — The Asiatic part of Turkey; the name sometimes designates all of Asia Minor. The excavation sites of Hacilar and Çatal Hüyük are situated in this plateau.

BYBLOS (bib′ lōs) — The earliest city of the Phoenicians and the site of excavations that yielded dark-faced burnished and combed pottery of the Neolithic period.

ÇATAL HÜYÜK (kə täl′ hú yúk′) — Most recent site of important Neolithic excavations in the foothills of the Taurus Mountains about 50 miles inland from the southeast coast of Turkey in Asia, i.e., Anatolia.

DALMA TEPE (dal ma′ tə′ pə) — A site in Iran.

DIMINI (di mi′ ni) — Site in Thessaly.

ERIDU (ā′ ri dü) — First royal city of the Sumerians located in present Iraq.

EYNAN (ā′ n ən) — Open site in Palestine disclosing Natufian industries.

HACILAR (hazh′ i lär) — Site of important Neolithic excavations in the Taurus foothills in Turkey near Çatal Hüyük.

HAJJI FIRUZ TEPE (hazh′ i fə rüs′ tə′ pə) — Excavation site in Persia, present-day Iran.

HASSUNA (ha sü′ na) — Site of excavations in northern Iraq.

JARMO (jär′ mō) — Site of excavations in the foothills of the Zagros Mountains of Iraq.

JERICHO (jer′ i kō) — Ancient site is a mound on the outskirts of modern Jericho in the Jordan Valley.

KHIROKITIA (kī rō kē′ tēə) — Excavation site on the island of Cyprus.

KNOSSUS (no′ sús) — Ancient city of Crete.

LASCAUX (läs kō′) — Site of Paleolithic cave paintings of the Magdelonian period in southern France.

MAGHOULA (mə gü′ lə) — A region in Thessaly.

MERSIN (mər sin′) — Situated in the Syro-Cilicia area of southeast Asia Minor, present-day Lebanon.

MESOPOTAMIA (mes′ ō pō tā′ mē ə) — The valley that lies between the Tigris and Euphrates Rivers. Modern Iraq.

MICHELSBERG (mi′ kəls bərg) — A culture of Bohemia, Germany and Switzerland.

RACHMANI (räk mä′ ne) — A site in Thessaly.

SESKLO (səs′ klō) — A site in Thessaly.

SOUFLI (sü′ flē) — A site in Thessaly.

SPIENNES (spān′ əs) — Site of flint mines in Belgium.

TEPE GURAN (tə′ pə gü ran′) — A site in Iran.

THESSALY (the′ sə lē) — Region of Greece.

ZAGROS (zä′ grōs) — Mountain range on the border between Iraq and Iran.

FOR FURTHER READING

Baldwin, Gordon C., *The World of Prehistory*. Putnam, 1963
Borer, Mary Cathcart, *Mankind in the Making*. Frederick Warne & Co. Ltd., 1962
Braidwood, Robert J., *Prehistoric Men,* 5th ed. Oceana, 1961
Braidwood, Robert J., and Gordon R. Willey, eds., *Toward Urban Life*. Aldine, 1962
Childe, V. Gordon, *What Happened in History*. Max Parrish, 1960
Coon, Carleton S., *The Story of Man*. Knopf, 1962
Daniel, Glyn, *The Idea of Prehistory*. World, 1963
Dickinson, Alice, *First Book of Stoneage Man*. Watts, 1962
Gabel, Creighton, ed., *Man Before History*. Prentice-Hall, 1964
Hawkes, Jacquetta, and Leonard Woolley, *Prehistory and the Beginnings of Civilization* (History of Mankind, Vol.1). Harper, 1962
Lucas, Jannette M., *Man's First Million Years*. Harcourt, Brace, 1941
Lynch, Patrick, *From the Cave to the City*. St. Martin's Press, 1959
Mellaart, James, *Earliest Civilizations of the Near East*. McGraw-Hill, 1965
Pfieffer, John, *Man's First Revolution*. *Horizon* Magazine, September, 1962
Quennell, Marjorie, and C. H. B. Quennell, *Everyday Life in the New Stone, Bronze and Early Iron Ages*. Putnam, 1923

Reference sources: Barnett, Lincoln, *The Epic of Man,* Golden Press, Time-Life, 1962; Beals, Ralph L., *An Introduction to Anthropology,* Macmillan, 1959; Braidwood, Robert J., *The Near East and the Foundations for Civilization,* an essay, Condon Lectures Oregon State System of Higher Education, 1962; Burkitt, Miles, *The Old Stone Age,* 4th ed., Atheneum paperback, 1965; *The Cambridge Ancient History—Anatolia before c. 4000 B.C. and c. 2300-1750 B.C.,* James Mellaart, rev. ed. V 1&2 Cambridge U. Press pamphlet; Childe, V. Gordon, *New Light on the Most Ancient East,* Grove Press, Evergreen paperback, n.d.; Coon, Carleton S. *The Story of Man,* Knopf, 1962; Frankfort, Henri, *Birth of Civilization in the Near East,* Doubleday Anchor book, 1950; Horgarth, Paul, and Jean Jacques Salomon, *Prehistory, Civilizations Before Writing,* V 6, paperback Dell, 1963; Hyma, Albert, *Ancient History,* College Outline Series, Barnes and Noble, paperback, 1940; Kenyon, Kathleen M., *Digging Up Jericho,* Praeger, 1957; Lloyd, Seton, *Early Anatolia,* Penguin Books, 1956; Lee, Norman E. *Harvests and Harvesting Through the Ages,* Cambridge University Press. 1960; Mellaart, James, *Earliest Civilizations of the Near East,* McGraw Hill, 1965; Mellart, James, "Roots in the Soil: The Beginning of Village and Urban Life," in *Dawn of Civilization* ed. by Stuart Piggot, McGraw-Hill, 1961. Pfieffer, John, "Man's First Revolution." *Horizon* magazine, September, 1962; Singer, Charles, et al *A History of Technology,* V 1, *From Early Times to the Fall of Ancient Empires,* Oxford, 1954; Titiev, Mischa, *The Science of Man,* rev. ed., Holt, Rinehart and Winston, 1963.

INDEX

This is a partially annotated index. Where no annotation occurs, the text or caption is fully explanatory. Italic numerals designate page reference to pictures.

Adze — a cutting tool having an arching blade set at right angles to handle, 22, *22*
Africa, 13, 24, *24*
Ain Sakhri, 46, *46*
Almonds, 13
Altamira, 46, *46*, 48
Amulets, 35, 46, 53
Anatolia, *see* Çatal Hüyük, Hacilar
Animals, 8, *9*, *9*, *10*, 11, 13, 15, 16, 18, *18*, 19, *19*, 20, *20*, 32, 34, *34*, 36, 37, 46, *46* 48, *49* 50, *50*, 51, *51*, 53; *see also* Bone(s); Bone implements; Bone ornaments; and specific names, i.e. "Sheep"
Anthropologists — specialists in the science of man in relation to physical character, distribution, origin, classification and relationships of races, environmental and social relations, and culture
Apples, 13
Apricots, 13
Archaeologists — specialists in the scientific study of the material remains of past human life and activities
Architecture, 26, *26*, 27, *27*, 28, *28*, 29, *51*, 52; *see also* Çatal Hüyük; Houses; Jericho; Lake dwellings; Post and lintel; Shrines; Stone; Tower
Art: Painting, cave, 46, *46*, *50*; painting, murals, Neolithic, 46, 48, *48*, *49*; *sculpture*, 46, *46*, 47, *47*, 50, *50*, lothic, 46, 48, *48*, *49*; *sculpture*, 46, *46*, 47, *47*, 50, *50*, 51, *51*, *52*, 53; *see also* Ain Sakhri; Amulets; Baskets; Bone; Çatal Hüyük; Clay; Stone
Asia, 18
Astronomers — specialists in the science which treats of the celestial bodies, 9
Aurochs, 18, *18*, 20, *20*, 48, *48*
Awl — a pointed tool for piercing small holes, 23, *23*
Ax, 22, *22*, 24, *24*

Barley, 11, *12*, 13, *13*, 14, 17
Baskets, 17, 30, *30*, 31, *31*, 38, 39, *39*; methods of making, 30; *30*
Beans, 14
Biologists — specialists in the science of living organisms, 9
Bone implements, 18, *18*, 22, *22*, 23, 25, 46
Bone ornaments, 18, *18*, 35, *35*
Bones, 9, 20, *20*, 22, 32, 53
Bricks, 26, 27, *27*
Butter, 19
Byblos, 45, *45*

Çatal Hüyük, 20, *20*, 22, *22*, 28, *28*, 32, *32*, 33, *33*, 35, *35*, 40, *40*, 46, *46*, 47, *47*, 48, *48*, *49* 50, *50*, 51, *51*, *52*, 53
Cattle, 18, *18*, 20
Cave dwelling, 26, *26*
Central America, 13, 54
Cheese, 19

China, 13
Clay, 30, 39, *39*, 40, *40*, 41, *41*, 43, 44, *44*, 45, *45*, 46, 53; firing, 40, *40*; pottery, 40, *41*, *42*, *43*, *44*, *45*, 46; preparation, 40, *40*; shaping, 40, *40*, 41, *41*
Clothing, 17, 19, 34, *34*, 35; *see also* Animals; Plants, use of; Weaving
Communities, 27, *27*, 28, *28*, 54
Cotton, 32, 34, *34*

Dalma Tepe, 23, *23*
Dates, 13
Digging stick, 15, *15*
Dog(s), 18, *18*, 19, 32

Egypt, 23, *23*, 24, *24*, 30, *30*
England, 25, *25*
Eridu, 39, *39*
Eynan, 46, *46*

Farmer(s), 8, *8*, 11, *11*, 13, 14, *14*, 15, *15*, 16, *16*, 28, 32, 36, 46, 47, 51, 53, 54; *see also* Women
Felt, 35
Figs, 13
Flaking, 22
Flax, 17
Flint — a variety of quartz such as chert, 22, *22*, 25, *25*
Flocks, 20, *21*, 28
Food: butter and cheese, 19; fruits, 13, 14; grains, 11, 12, 13, 14, 15, 16, 17; meat, 14, 18, 19; milk, 18, 19; nuts, 13, 14; vegetables, 13, 14, 17; *see also* specific names, i.e. "Honey"
Fruits, 14; *see also* specific names, i.e. "Apples"
Funeral ceremony, *52*, 53

Geologists — specialists in the science of the history of the earth and its life, especially as recorded in the rocks, 9
Goats, 19, *19*, 20, *21*, 32
Gourds, 17, *17*
Grains, 11, *11*, 13, 15, *15*, 16, *16*, 17, *17*; *see also* Barley; Seed(s); Wheat
Grand Passigny, France, 25, *25*
Grasses, 30, *31*

Hacilar, 22, *22*, 35, *35*, 43, *43*, 44, *44*, 46, *46*, 47, *47*
Hajji Firuz Tepe, 42, *42*
Harpoon, 23, *23*
Hassuna, 22, *22*, 45, *45*
Hemp, 17, 32
Herds, 20, *21*, 28, 51
Hoe, 22, *22*, 24, *24*
Honey, 14

Horse, 18, *18*, 22
Houses, 9, 26, *26*, 27, *27*, 28, *28*; *see also* Cave dwelling; Lake dwellings; Long house; Mud brick; Round houses; "Tauf"; Wattle and daub; Windbreak
Hunter(s), *10*, 11, 20, 22, 30, 34, 36, 46, 48, *49*, 50, *50*, 53

Indus Basin, 54
Iran, *12*, 13, 23, *23*
Iraq, *12*, 13, 39, *39*

Jarmo, 12, *12*, 20, *20*, 23, *23*, 27
Jericho, 22, *22*, 23, *23*, 26, 27, 29, *29*, 35, *35*, 53, *53*

Khirokitia, 35, *35*
Knossus, 41, *41*

Lake dwellings, 22, *22*, 27, *27*
Lascaux, 48
Long house, 27, *27*

Maghoula, 35, *35*
Matting, 17; techniques, 30, *31*
Meat, 18
Mediterranean, *12, 13*, 54
Mellaart, James, 48
Mersin, 45, *45*
Mesopotamia, 23, *23*
Michelsberg, 41, *41*
Milk, 18, *19*
Mines, 25, *25*

Naturalists — specialists in the science that treats of animals and plants, 9
Neolithic, 9, 20, 22, *22*, 23, *23*, 28, 29, 38, 40, *40*,
Neolothic, 9, 20, 22, *22*, 23, *23*, 28, 29, 38, 40, *40*,
41, 42, 44, *44*, 45, *45*, 47, *47*, 53, *53*, 54
Nile Valley, *12*, 54
Nuts, 14; *see also* specific names, i.e., "Almonds"

Oats, 14
Olives, 13
Ornaments, 18, 35, *35*, 46, 53; *see also* Amulets; Bone ornaments; Stone

Peaches, 13
Pears, 13
Pick, 25, *25*
Pig(s), 15, 18, *18*, 20
Plants, use of, 16, 17, *17*, 26, *26*, 30, *30*, *31*, 32, 34; *see also* specific names, i.e., "Cotton"
Plums, 13
Post and lintel, 29, *29*
Pottery, *see* Clay
Pumpkins, 17, *17*

Quern (kwûrn) — a primitive hand mill for grinding grain, 23, *23*
Quffa (kufi fá), 39, *39*

Rachmani, 43, *43*
Rake, 25, *25*
Reeds, 30, 31, *31*, 38, *38*
Religion, 48, *48*, 50, *50*, 51, *51*, 52, 53; *see also* Funeral ceremony
Round houses, domed stone, 26, *26*
Rushes, 30, *31*
Rye, 14

Seed(s), 9, 11, 14, *14*, 15, *15*, 16, *16*; *see also*, Barley; Grain(s); Wheat
Shards, 40, *40*, 44, *44*
Sheep, 15, 19, *19*, 20, *21*, 32
Shovel, 25, *25*
Shrines, 28, *28*, 50, *50*, 51, *51*,*52, 53; *see also* Çatal Hüyük
Sickle, 22, *22*, 23, *23*
Skin(s), animal, *10*, 19, 20, 37, 38, 39, *39*; tanning, 34
Soufli, 35, *35*
South America, 13; Peru, 54
Spiennes, Belgium, 25, *25*
Squashes, 17, *17*
Stone, 15, *15*, 17, *17*, 22, *22*, 23, *23*, 24, *24*, 25, *25*, 26, 27, 29, *29*, 35, *35*, 46, *46*, 47, *47*, 50, *50*, 51, *51*, 53; shaping and polishing, 22, 24; *see also* Flaking; Flint; Tools; Utensils

"Tauf"—coloquial arabic, 27
Tepe Guran, 44, *44*
Thessaly, 35, *35*, 43, *43*, 44, *44*
Tools, 9, *11*, 15, *15*, 16, *16*, 22, *22*, 23, *23*, 24, *24*, 25, *25*; *see also* Bone implements; Flint, Stone
Tower, Jericho, 29, *29*
Transportation — Land, animal skin, 37, *37*, pole support, 36, *36*; sledge, 37, *37*; tump line, 36, *36*
Transportation — Water, clay boat, 39, *39*; *coracle*, 39, *39*; dugout, 38, *38*; quffa, 39, *39*; raft, 38, *39*

Utensils, 17, *17*, 19, 20, 23, 40, *40*, 41, *41*, 42, *42*, 43, *44, 45, 46*; *see also* Clay; Stone

Wattle and daub, 26, *26*, 27, *27*
Weapons, 9, *10*, 20
Weaving, 17, 19, 30, *30*, *31*, 32, *32*, 33, *33*, 34, 46; frame, 32, 33, *33*; *see also* Baskets, Çatal Hüyük; Matting; Spindle; Whorl
Wheat, 11, *12*, 13, *13*, 14, 17
Whorl — small flywheel of a spindle, 32, *32*
Windbreak, 26, *26*
Wolf, 18
Women, 11, *11*, 14, *14*, 15, 16, 17, 31, 32, 33, 34, 35, 40, 47, 50, 51
Wood, 22, *22*, 27, *27*
Wool, 19, *19*, 32, 34, *34*, 35

Zagros Mountains, *12*, 13, 54

Leonard Weisgard, the author-artist of THE FIRST FARMERS, has illustrated more than 200 books for children. Among them is an earlier *Life Long Ago* book, *The Athenians*. In THE FIRST FARMERS he has turned his wide and varied talents to a re-creation of the world of New Stone Age man. In vital, detailed pictures he shows how Neolithic man worked, farmed, and made useful and beautiful tools and utensils, some of which still exist. To be sure that every detail is completely authentic, all the artifacts shown are redrawings of actual finds selected by Mr. Weisgard from many scholarly sources.

Mr. Weisgard has had, throughout, the assistance of Edward Ochsenschlager, Historical and Editorial Consultant for the *Life Long Ago* series. Mr. Ochsenschlager is an archaeologist and a lecturer in classics at Brooklyn College. He is Associate Editor of *The Classical World*. In 1961 and 1962 he was the Assistant Field Director of the New York University Anatolian Expedition at Aphrodisias. Currently he is the Classical Archaeologist at the Mendes excavation, sponsored by the Institute of Fine Arts of New York University, the Brooklyn Museum, the Detroit Museum of Art, and the Oriental Institute of the University of Chicago. He has reviewed both text and illustrations for historical accuracy and has contributed a list and map of selected, representative, important Neolithic sites.

Rosemary Daly, Education Consultant for the *Life Long Ago* series, has helped plan the entire book to make it of maximum use to readers and has prepared the glossary and annotated index. Miss Daly is Librarian of the Ethical Culture School in New York City.

Books in the Life Long Ago Series

The Life Long Ago books are close-up views of ancient civilizations. Everyday life is brilliantly re-created in panoramic scenes, authentic detailed drawings, and concise text. Each book is a rare visual experience. Each takes the reader into the reality and excitement of history and provides an extraordinary understanding of a people and their ways.

THE FIRST FARMERS in the New Stone Age

Pictures and Text by LEONARD WEISGARD

This is the story of one of the greatest revolutions in the life history of man—the discovery of agriculture. You see how man lived in the first settled communities. You watch the earliest farmers at work—sowing grain, tending their herds, making tools, and developing the crafts of pottery, weaving, and basketry. A striking pictorial treatment of life in the New Stone Age.

THE CAVE DWELLERS in the Old Stone Age

Pictures and Text by RICHARD M. POWERS

The pictures burst with life and strength. You feel the tense anticipation of the hunters as they gather before the magician in the ritual cave. You know the fear as they face a live mammoth, and realize they must kill or be killed.

THE EGYPTIANS in the Middle Kingdom

Pictures by SHANE MILLER
Text by EDWARD OCHSENSCHLAGER

A fascinating trip through this ancient land where you will visit the lonely Pharaoh of Egypt, walk through the shop-lined streets of Memphis, and journey past the great pyramids of Giza.

THE ROMANS in the Days of the Empire

Pictures and Text by SHANE MILLER

The mighty Roman Empire is vividly alive in the strength of these pictures. You take your seat among 45,000 spectators at the Colosseum. You follow a freedman home to his four-story apartment house and you feel the might of the Roman army as they prepare for battle.

THE ATHENIANS in the Classical Period

Pictures and Text by LEONARD WEISGARD

The beauty and reality of Athens is before you in the streets of the city as you visit an instrument maker's shop, join the crowds at the Panathenaic Stadium and stand before the Parthenon.

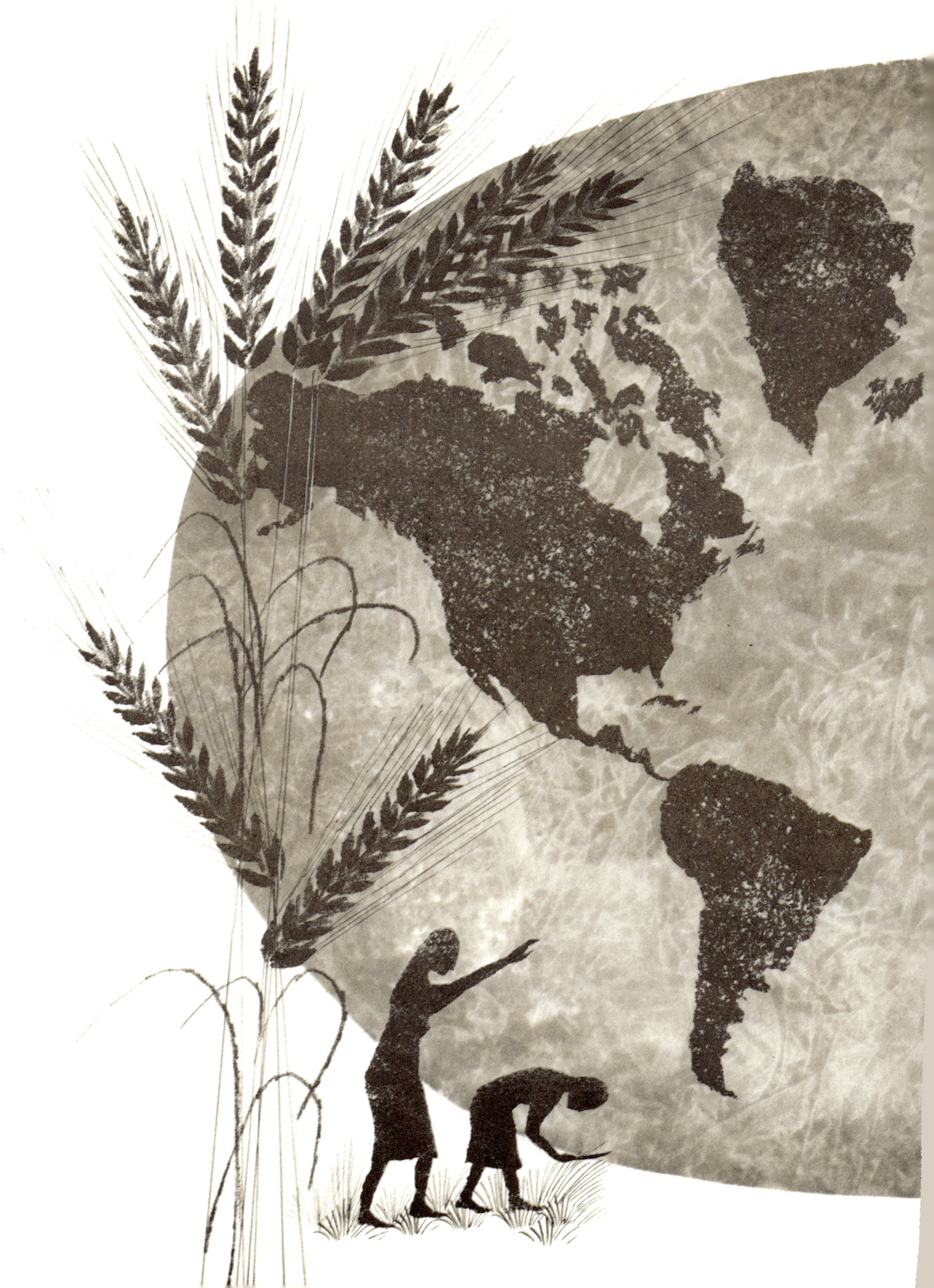

EARLIEST NEOLITHIC WORLD